Climate Basics

Climate Basics series

Climate Basics: Nothing to Fear, by Rod Martin, Jr.—an Amazon #1 Bestseller in Weather and Science & Math Short Reads

Deserts & Droughts: How Does Land Ever Get Water? by Rod Martin, Jr.

Shining a Light Series

Dirt Ordinary: Shining a Light on Conspiracies, by Rod Martin, Jr.

Favorable Incompetence: Shining a Light on 9/11, by Rod Martin, Jr.

Thermophobia: Shining a Light on Global Warming, by Rod Martin, Jr.

Climate Basics
Nothing to Fear

Rod Martin, Jr.

Tharsis Highlands Publishing
Cebu, Philippines

Published by Tharsis Highlands Publishing
Cebu, Philippines
https://tharsishighlands.wordpress.com/books/

Amazon Print Edition
May 2019
ISBN: 9781095733004

EBook Editions
Amazon Kindle—2018, 2019
Smashwords—2018, 2019

Cover photos(details, posterized): Desert sunset by S.saban (CC BY-SA 4.0); Grassland, Cantabria, Spain by Serdio (CC BY-SA 3.0); Harvest, Brandon, Manitoba (PD); Lake Fryxell, Transantarctic Mountains by Joe Mastroianni, National Science Foundation (PD). Cover design: Rod Martin, Jr.

Typography fonts
Headings: Rockwell Extra Bold
Running Heads: Rockwell
Text: Palatino Linotype

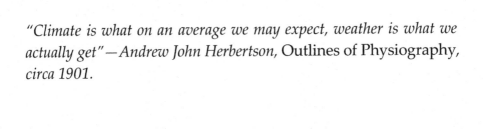

"Climate is what on an average we may expect, weather is what we actually get" — *Andrew John Herbertson,* Outlines of Physiography, *circa 1901.*

Table of Contents

Introduction: Simple Basics

Most of us care about this planet and the life that lives on it. This is our home. A few with power don't seem to care, and that is a shame. They threaten the future for the rest of us.

In order to find the best policies for our mutual future, we need to look at the pros and cons of every idea. We need to weigh each of them in the context of the current reality. And, on the subject of context, we need to remain precise in our description of the current situation and the factors which affect the changes we are currently undergoing. Without that precision, we cannot accurately assess the situation, and have no hope of finding the right solution.

Here is what I have found....

Right now, we live near the peak of a major Holocene warm period (see Glossary), 1,000-year cycle, in a moderately cool interglacial, in the midst of a 2.6-million-year Ice Age. This may sound quite strange to those who have been listening only to the alarming reports of Earth's so-called "fever." Ironically, our current warm period is the coldest of the Holocene's ten major warm periods.

Having studied climate science starting in the mid-70s, I've long known a few things on the topic. I've also long believed in taking responsibility for self, family, nation, environment and the world. So, it should come as no surprise that I took Al Gore's original film to heart when I first saw it. In 2012, I woke up to the fact that Al Gore and I had been horribly wrong.

How did I come to this conclusion? The answer is quite simple. I had returned to the basics of climate science. Because of this, I started to see one logical fallacy after another in Gore's narrative, in that of the IPCC and in the story being woven by the corporate news media.

Some people have the bizarre notion that climate change is new. It isn't. In fact, climate has done nothing but change ever since Earth gained an atmosphere, more than four billion years ago.

This short book covers only the basics, taking one lie after another and setting the record straight. Here's what is in store for you:

Chapter 1—Debunking the "more storms" idea.

Chapter 2—Debunking the "more droughts" belief.

Chapter 3—Debunking the notion that global warming is bad.

Chapter 4—Debunking the concept that CO_2 drives global warming.

Chapter 5—Revealing the fact that cooling in our ongoing Ice Age is dangerous.

Chapter 6—Debunking the idea that we are anywhere near dangerous levels of CO_2.

Chapter 7—Revealing the far more likely climate forecast.

If you want more in-depth information on climate, with references and sources, you might consider two of my other books on the topic:

- *Thermophobia: Shining a Light on Global Warming*
- *Red Line — Carbon Dioxide: How humans saved all life on Earth by burning fossil fuels*

And you can find out more about these books at, **https://TharsisHighlands.WordPress.com/books/**

For additional information on climate, see the Global Warmth blog at, **https://globalwarmthblog.wordpress.com/blog/**

Note on Illustrations

To view the illustrations shown in this book both full-size and in color, go to the following internet address: **https://tharsishighlands.wordpress.com/2018/11/18/illustrations-used-in-climate-basics/**

Chapter 1: Wind

Warming alarmists are fond of telling us that global warming will result in more violent storms—not only a far greater number of storms, but far stronger ones, too.

Well, this is only partly true.

We have to ask ourselves: What causes wind to blow?

It's not heat. Venus has plenty of heat, but virtually zero wind at the surface. Million-year-old craters have remained in pristine condition because of that lack of wind and the erosion that might otherwise have degraded the form of those craters. If you think Venus has strong winds at the surface, you're wrong. The strong winds on our sister planet are above the Venusian clouds, more than 50 kilometers above the surface.

Jupiter is deathly cold—colder than the most frigid spot on Earth—but has the largest storms in the solar system. The Great Red Spot storm is larger than our entire planet.

Wind? Heat alone couldn't cause it to blow. With the same hot temperature everywhere on Venus—approximately 462°C—wind has no reason to blow, and it would have no direction to blow. Wind blows because of temperature differences.

We're all familiar with the idea that hot air rises. Smoke rises from an open fire. Hot air balloons rise into the sky. And we're also likely familiar with the concept that cold air descends. If you've ever been to a mall on a hot day, you may have noticed the cold air rushing by your ankles as the mall doors open for you.

Key: Wind blows because of temperature differences

Some may think that pressure differences are what cause wind to blow, but that's only the middle of the story. Temperature is what controls pressure. Hot air expands, causing high pressure. Cold air contracts, causing low pressure. See? The simple basics on wind point to temperature differences as the culprit.

But what does this have to do with the count and strength of violent storms?

Warmer oceans will result in hurricanes becoming stronger, because those cool winds will be passing over warmer waters, generating more energy. It's kind of like a battery. The greater the voltage (sometimes called "potential"), the brighter your light burns. Temperature difference is like the voltage in a battery.

How the warming alarmist meme is wrong involves the planet as a whole. When you warm up our Earth, melting all that polar ice, you end up reducing the overall temperature difference on the planet.

Hurricanes have shown at least 45 years of decreasing counts, but increasing strength. Eliminating polar ice could eliminate hurricanes altogether. So, it wouldn't matter how warm the oceans became, from the standpoint of storms, because zero cyclones means zero level of violence. Counts of violent tornadoes in the United States (EF3–EF5) are also down—a 60-year downtrend.

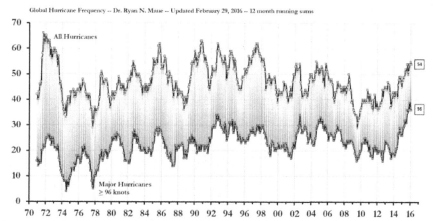

Hurricanes (including typhoons) 1971–2016, from Dr. Ryan Maue. Notice the overall slightly downward trend in hurricane counts. Also notice that the count of major hurricanes is slightly up. This remains consistent with the idea that wind only blows because of temperature differences. As the Earth warms, the count reduces toward zero while the overall energy of hurricanes goes up. Eliminate all polar ice and we might have very few hurricanes—perhaps even zero. To view the figure full-size and in color, see internet address at the end of the Introduction.

Dr. Hubert Lamb, founder and first director of the Climate Research Unit at East Anglia University in the UK, researched violent storms during the Little Ice Age. He found that there were a far larger count of violent storms each century during the depths of that cold climate than during the entire 20th century.

You may have heard of one of those storms. The year was 1588 and the English were terrified they might have to brush up on their Spanish. But the great storm that year sank most of the Spanish Armada. Another one—the Great Storm of 1703—wracked the English coastline with hurricane force winds. Imagine that: a hurricane in England!

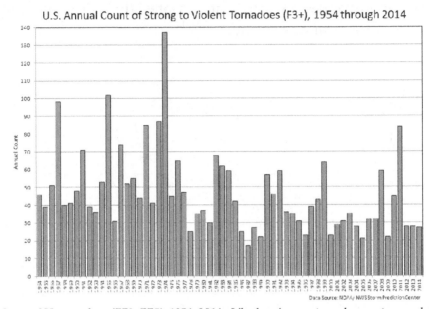

U.S. Annual Count of Strong to Violent Tornadoes (F3+), 1954 through 2014

Strong US tornadoes (EF3–EF5) 1954–2014. Like hurricanes, tornado counts are also down. Unlike hurricanes, tornadoes don't normally have the boost from cool winds passing over warmer waters. As polar ice is eliminated, tornadoes become increasingly rare. To view the figure full-size and in color, see internet address at the end of the Introduction.

Now that we know that wind only blows because of temperature differences, we know why there were more violent storms. When polar cold moves closer to equatorial heat, the temperature gradient becomes steeper, jacking up the thermal potential.

Conclusion: Global warming reduces violent storms

Chapter 2: Rain

Warming alarmists are fond of telling us that global warming will result in more droughts and deserts. This makes sense until you dig deeper and ask: How does land ever get water in the first place?

If you reduce global temperature, you get less rain and far more droughts and deserts. Why? Because rain starts with warmth and evaporation. If there were no warmth, then land would never get any rain and it would all be desert with zero life.

We've all seen how a hot day will evaporate a puddle of water far faster than a cool, overcast day. But clouds would never form if we didn't have evaporation. The more water that evaporates from the ocean, the more rain we will have. Simple. But I've had people argue with me that warmer air will be able to hold more water, so the greater evaporation won't cause more rain. How this is horribly wrong is easy to see with a little simple arithmetic.

Key: Rain starts with warmth and evaporation

Let's say you have 2 units of evaporation to start with, and 2 units of rain each day. If climate were to warm up so that you had 3 units of evaporation, you would soon have 3 units of rain each day.

But let's entertain the notion that more evaporation would not lead to more rain. After one week, you would have 21 units of evaporation at our new, higher rate, and only 14 units of rain. That would leave 7 units of water in the sky. At the end of a year, you would have 365 units of water in the sky. At the end of a century, you would end up with 36,500 units of water above our heads. And if this continued unabated, you would eventually end up with all of the oceans above our heads. See?

Warmer air will be able to hold more water, but once that has reached saturation, any additional evaporation will result in more rain.

When you listen to the news and hear of droughts and floods, remember that the corporations need profits and news makes profits by making the news seem alarming. The events about which you hear are anecdotal. They really happen, but they aren't the whole picture.

During the far warmer Holocene Optimum, the Sahara was green for 3,000 years. In fact, there was so much extra rain during that far warmer period, that today's wimpy Lake Chad was, back then, a far larger, robust inland sea which rivaled the Caspian in its heyday. When Earth cooled off, the climate became drier and people escaped to the Nile River Valley, triggering the start of civilization.

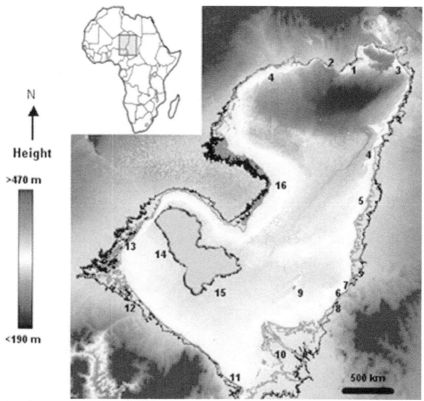

Map showing the outline of Lake Megachad at its maximum extent during the Holocene Optimum.   Image   by   Nick   Drake,   courtesy   of   King's   College. https://kcl.ac.uk/sspp/departments/geography/people/academic/drake/Research/The-Sahara-Megalakes-Project/Lake-Megachad.aspx

We will always have droughts and deserts, but global warming will reduce both across the planet.

Conclusion: Global warming reduces droughts and deserts

Chapter 3: Beneficial Warmth

Warming alarmists are fond of telling us that the world is going to burn up. NASA, which used to deal with science, has now started to appeal to emotion (a common logical fallacy) with such statements as, "Earth has a fever." Funny how the identical logically fallacious line appeared at the same time in a silly spy movie. They imply that global warming is dangerous.

It should be obvious that life prefers warmth to the cold. A quick survey of population density by latitude shows this. The population density of Greenland is close to zero people per square kilometer, while the population density of Java is over 1,000 people per square kilometer.

Naturally, warmth is not the only criterion. Life also needs water. The polar regions are deserts just as much as is the Sahara, because both regions get very little precipitation each year.

The key reason that global warming is good is that we still live in an Ice Age. This may come as a shock to some people. Clearly, there is some confusion on this and differing opinions. But look at the following facts:

- The Pleistocene (2.6 million to 11,600 years ago) is frequently called the most recent Ice Age.
- The Pleistocene consisted of dozens of glacial and interglacial periods.
- The Holocene (our current period) is an interglacial.
- The Holocene is, by far, not the warmest interglacial of the Pleistocene.

Key: We still live in an Ice Age

Any reasonable person would look at the above facts and realize that we're still in the Pleistocene Ice Age, despite the artificial claim that it ended 11,600 years ago. When that idea was originally presented (Pleistocene ended), we knew very little about our climate history. We knew there had been a period of extreme cold and that we were no longer in that extreme cold. What people a century ago didn't know is that the amount of cold is merely relative; our current warmth is still very cold compared to the epoch before our current Ice Age began.

In fact, we currently live near the bottom of Earth's livable temperature range. Sure, too much warmth would be a bad thing, but we're far from suffering that problem. Too much of anything would be bad. Saying that we have more of something than we've had in thousands or millions of years isn't saying much. We need to know this information in the proper context.

The proper context

The corporate news media is owned by big corporations which, in turn, are owned by globalists who have their own, not-so-secret agenda. So many people do not trust big corporations, yet never give it a second thought to trust the evening news. And that's a problem.

Climate always changes and always has. We either get warming or cooling. And climate changes in cycles. Right now, we're in the midst of the Modern Warm Period—part of a 1,000-year cycle of major Holocene warm periods. We're also in an interglacial of a 2.6-million-year Ice Age. According to the work of W.S. Broecker (1998), interglacials of the last million years or so have lasted an average of 11,000 years. My own research has determined that the range of interglacial durations has been from 4,000 to 28,000 years. Our current interglacial—the Holocene—is as much as 17,000 years old, if we include the massive warming before the 1,300-year Younger Dryas "Big Freeze."

Some people have argued that it's not the amount of warming, but the rate of warming that is dangerous. The warming alarmists have repeatedly quoted 3°C as the likeliest amount of global warming over the 21st century. That amounts to 0.00008219°C of warming each day, on average. Such a tiny rate of warming is near impossible to measure and impossible to feel. No one is going to be keeling over with a "blistering" rate of warming like that.

Throughout history, warming has led to prosperity and abundance. Civilizations have found it easier to expand and to build.

- Minoan Warm Period (1400–1100 BC)
- Roman Warm Period (200 BC–100 AD)
- Medieval Warm Period (850–1350 AD)
- Modern Warm Period (1850–Now)

And throughout history, cooling has led to famines and societal collapses. Some civilizations were able to hold on, but cooler climate made it more difficult. For instance, the Notre Dame building project started just before the end of the Medieval Warm Period, while French prosperity made it possible to embark on such an ambitious development. The massive cooling at the start of the Little Ice Age threatened to leave the project unfinished.

The 1,000-year cooling cycle includes:

- The Greek Dark Ages (1100–800 BC)
- The Post-Roman (Medieval) Dark Ages (500–850 AD)
- The Little Ice Age (1350–1850 AD)

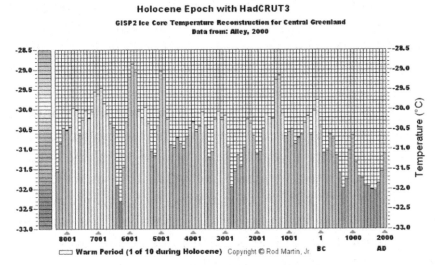

This is a graph I developed from GIPS2 data using color coding to dramatize the cooling trend of the last 3,000 years. Warming alarmists imply that the Modern Warm Period is the hottest period of the Holocene. Clearly this is not the case. The Medieval Warm Period was the second coldest warm period and Vikings could grow crops in Greenland for hundreds of years. Today, Greenland has not yet reached the Viking standard. To view the figure full-size and in color, see internet address at the end of the Introduction.

Nothing in climate is clear cut or hard edged. The start of one period isn't always a stark change from the previous day. Cooling trends may have warming reversals, and vice versa. One region may get more rain because of warming, while another region may receive less rain. But warming is good for the planet as a whole. More places receive life-giving rain, because there is more evaporation from the oceans. Flooding has always happened, and there is no evidence that flooding becomes more frequent. One study looked at 500 years of floods in Europe and found no significant trend up or down in the number of floods or their strength.

As we've seen from scientific evidence, droughts become less frequent and less severe, while deserts shrink in size, but become harder (drier in their interiors).

There is a big reason why human population could not get above 100,000 during the last glacial period that lasted some 90,000 years. Cold oceans don't evaporate enough water to get rain to the land for agriculture. With the massive warming that kicked off the Holocene, rain was suddenly abundant enough and predictable enough for humans to start agriculture.

Conclusion: Global warming in our ongoing Ice Age is a good thing

Chapter 4: Causes of Warmth

Warming alarmists are fond of telling us that global warming is being driven by Man's output of CO_2. Even the United Nations' IPCC created Terms of Reference which specifically and only focus on human sources of warming. That's science with blinders on.

The vast majority of warmth that drives our climate comes from that "little" yellow ball in the daytime sky. Though the amount of light we receive from our sun doesn't change very much, the sun has one other method for affecting climate on Earth that is far more potent than most other factors. The other, more potent solar influence involves solar wind.

Some of the major climate factors that affect our planet's warmth include,

- Solar wind
- Milankovitch cycles
- Atlantic multi-decadal oscillation
- Pacific decadal oscillation
- El Niño and La Niña

The Atlantic and Pacific cycles involve ocean currents. Milankovitch cycles involve properties of Earth's orbit and tilt.

Humans have affected climate on the local level with things like urban heat island effect and land use changes (clearing, farming and other changes).

Key: The primary source of warmth is the sun

Solar wind was discovered as a driving force in Earth's climate by Henrik Svensmark and others. The more active is the sun, the more solar wind it emits, and this pushes against the cosmic rays bombarding the solar system from distant stars (supernovae). When there is more solar wind, there are fewer cosmic rays that make it to Earth, and fewer clouds formed from the nucleation process (cloud chamber effect). When there are fewer clouds, the Earth becomes warmer, because more sunlight makes it down to the surface. When the sun is relatively inactive, producing less solar wind, more cosmic rays make it to Earth's atmosphere, stimulating the formation of more clouds, reflecting more sunlight and cooling down the planet. This one effect has been shown to produce a far greater correlation with global temperature than most other factors.

Carbon dioxide, on the other hand, shows virtually no correlation with temperature on nearly every time scale. On the one time scale with a strong correlation between CO_2 and temperature, temperature drives CO_2 abundance; not the other way around. Carbon dioxide is such a weak influence on temperature that, though CO_2 continued to increase in the paleoclimate record, temperatures fell despite the increasing carbon dioxide levels.

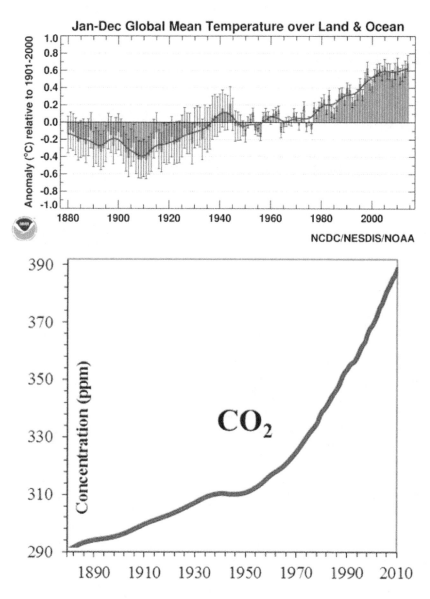

Temperatures from NOAA, 1880–2010, with error bars on degree of certainty. Note that graphs in the 1970s showed far steeper cooling from 1940. This was what gave us the "Ice Age" scare back then. The CO₂ graph was originally from 1750–2010. It has been trimmed and resized to match the NOAA graph time scale. Notice how CO₂ and temperature have poor correlation with temperature going up and down while CO₂ is only going up.

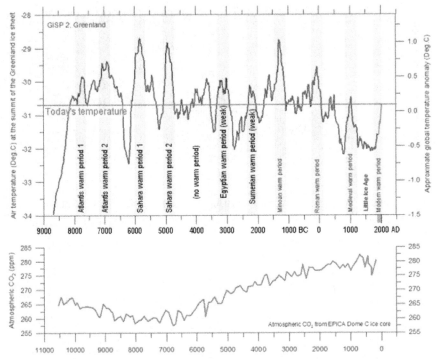

10,700 years of temperature and CO2. This composite graph includes temperature proxy readings from Greenland (GISP2 site) and CO_2 readings from EPICA Dome C (Antarctica) modified to match the time scale of the temperature graph. The original temperature graph had only 4 green bars. I added the additional green bars to show the periodicity of the major warming periods of the Holocene. I've also added the red lines for "Today's Temperature" and the extended temperature graph from the end of the ice core temperature proxy. Notice how CO_2 and temperature have very poor correlation on this time scale. Also, notice the strong cooling trend from the Minoan Warm Period (c.1100 BC). To view the figure full-size and in color, see internet address at the end of the Introduction.

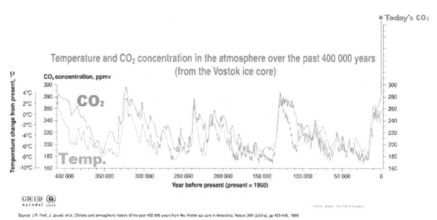

400,000 years of temperature and CO_2. This graph has been modified by placing both temperature and CO_2 together for closer comparison. I've also added the sharp modern rise in CO_2 to today's level. Notice that despite the tight correlation on this time scale, there is not a commensurate spike in temperature to match the massive spike in carbon dioxide. The strong correlation on this time scale between temperature and CO_2 is a function of temperature driving CO_2 in and out of the oceans. The lack of a commensurate spike in temperature for today shows that the effect of CO_2 on temperature is extremely weak, if it exists at all.

67 million years of temperature and CO_2. Notice how low our current temperature is on the far right of the graph. And notice how CO_2 and temperature have very poor correlation across most of the graph.

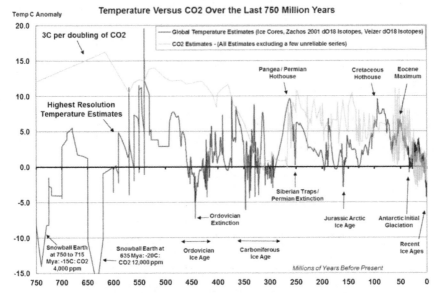

750 million years of temperature and CO2. Notice how temperature and carbon dioxide seem to have virtually no relationship at all. Without correlation, there can be no causation. To view the figure full-size and in color, see internet address at the end of the Introduction.

This point cannot be stressed enough: When there is virtually no correlation, there can be no causation. CO_2 does not drive global temperature. In fact, it has been shown to have virtually no effect on global warming on all time scales.

And the modern, massive spike in CO_2 from human activities has not produced a commensurate spike in temperature. Based on the behavior of CO_2 on the 10s to 100s of thousands of years scale, we would expect the modern boom in CO_2 levels to produce at least +10°C of warming, if CO_2 were driving temperatures. But that hasn't happened. In fact, the warming of the last 70 years has been less than 1°C. And virtually all of that has been from the natural, 1,000-year cycle of warming.

Conclusion: CO_2 has very little effect on global temperature

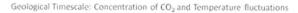

Geological Timescale: Concentration of CO_2 and Temperature fluctuations

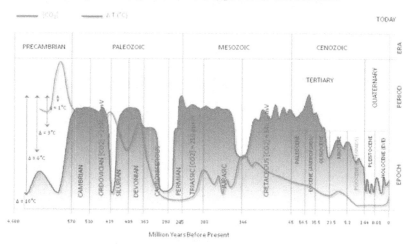

1 - *Analysis of the Temperature Oscillations in Geological Eras* by Dr. C. R. Scotese © 2002. 2 - Ruddiman, W. F. 2001. *Earth's Climate: past and future*. W. H. Freeman & Sons. New York, NY. 3 - Mark Pagani et all. *Marked Decline in Atmospheric Carbon Dioxide Concentrations During the Paleocene*. Science: Vol. 309, No. 5734: pp. 600-603. 22 July 2005. *Conclusion and Interpretation* by Nasif Nahle ©2005, 2007. *Corrected on 07 July 2008 (CO2, Ordovician Period)*.

4,500 million years of temperature and CO_2 (4½ billion years). I've added temperature shading to emphasize how cool it is today and how much most of Earth's history has remained far, far warmer than today. During most of that time, life thrived. Warmth, within this range of temperatures, is good for life. Notice again how temperature and CO_2 have virtually no correlation on this time scale. What little correlation there is remains accidental and not cause-and-effect. To view the figure full-size and in color, see internet address at the end of the Introduction.

Chapter 5: Dangerous Cooling

Warming alarmists are fond of ignoring the dangers of global cooling.

We live in an Ice Age. Those two "little" white things at the poles remind us of that fact. After all, they are made of ice. And our Holocene interglacial is cooler than many of the interglacials of the current Ice Age.

The previous interglacial was as much as +5°C warmer than now—and that, for hundreds of years. Polar bears did just fine, and sea level was as much as 20 meters higher than today.

Life struggles in the cold. Penguins know that if they leave their eggs for more than a few seconds, the chick inside will die. Polar bear numbers have been increasing throughout the Modern Warming Period. More warmth means more food.

Humans also suffer from cooling. As we've mentioned, cooling periods result in famines and societal collapses. The reason is quite simple: cold oceans generate less rain, making it more difficult to grow crops. People tend to get rowdy when they're starving to death.

Key: Cold kills

There is strong evidence that the Holocene started to shut down 3,000 years ago. We've been on a strong cooling trend since the height of the Minoan Warm Period. The drastic cooling which led to the Greek Dark Ages found temperatures about where they are today. Let that fact sink in.

Up and down, the global climate has warmed and cooled, but the 3,000-year trend has been ever downward. The Post-Roman, Medieval Cool Period (Dark Ages) was the second coldest period of the Holocene, not counting the Younger Dryas. The Medieval cold spell reached colder temperatures than the Little Ice Age, but not by much. And the Little Ice Age lasted far longer, giving Earth a far larger volume of cold than any other period since the Younger Dryas.

Of all the ten major warm periods, the Modern has been the coldest. The second coldest—the Medieval Warm Period—was warm enough to allow the Vikings to grow crops in Greenland for hundreds of years. So far, we're not up to Viking standards in Greenland.

According to the work of W.S. Broecker, our Holocene is already as much as 6,000 years older than the average interglacial. The second longest interglacial of the last million years was about 18,000 years in duration, and we're only 1,000 years shy of that, if we count the warmth before the Younger Dryas.

I've heard claims that Milankovitch cycles will give us another 12,000 to 50,000 years of additional warmth. But those estimates don't take into account solar wind cycles. There is still a great deal we do not know about climate and the triggers for glacial periods of the current Ice Age.

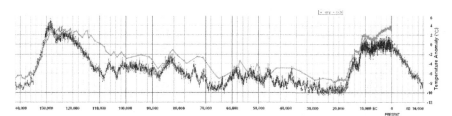

This is a detail from a far longer graph of temperature and CO_2 proxies from Antarctica, going back 800,000 years. This has been modified by adding about 10,000 years from now, into the future to show what an Eemian cool-down would look like from the current 3,000-year cooling trend. The Eemian was the previous interglacial of the current Ice Age. The warmest period of the Eemian peaked about 128,000 years ago. To view the figure full-size and in color, see internet address at the end of the Introduction.

With all that we do know, there is a very real danger that the Holocene will continue to shut down, leading into the next glacial period of the current Ice Age. With that brutal knowledge, your author only wishes CO_2 had more control over global temperature, but it doesn't.

Thankfully, interglacial periods cool down far more slowly than they warm up. Typically, the warm-up takes tens to hundreds of years; cooling takes hundreds to thousands of years. All other things being equal, the slide into glacial period cold would likely take another 5,000 years.

Ironically, the UN and other political entities want to cool down the planet artificially. Former CIA Director John Brennan praised Stratospheric Aerosol Injection (SAI) as one promising method (CFR, June 2016). He liked the idea that we could cool down our world "like volcanoes do." His statement generated a double helping of irony, because he said it on the 200th anniversary of the 1816 year without a summer—volcanic cooling that killed thousands and turned thousands more into climate refugees. The level of irresponsibility and ignorance on the part of Brennan and others remains one of the mind numbing insanities of our modern civilization.

Though the last glacial period likely never saw human population get above 100,000, the upcoming glacial period might see far more, because we now have technology to help us out. But how much of that technology will we be able to maintain? How will cooling climate, famines and human needs erode or fracture our technological infrastructure? Only time will tell.

Some warming alarmists invoke the "precautionary principle" to guide us through these times of abundance and fake news hysteria. But doing anything just to be doing something is the height of insanity. No one in their right mind would jump off a cliff just to be doing something. We need to use more caution, take better stock of the facts and choose wisely the correct direction.

Conclusion: We live in an Ice Age and global cooling is the real danger

Chapter 6: Beneficial CO$_2$

Warming alarmists are fond of telling us that CO$_2$ is a pollutant.

Carbon dioxide is not, and never has been, a pollutant. It is a vital gas of life without which all life (plants and animals) would die. Life can continue without oxygen—plants, that is. But if we don't have CO$_2$, all plants will die, then all animals will die, because they won't have any food.

CO$_2$ levels have been far, far higher in the past. When life first started, CO$_2$ levels were as much as 965,000 ppm. We currently have 210,000 ppm oxygen; there was a time when all that oxygen was coupled with carbon, so the atmosphere during the time of the earliest photosynthesis may have had around 210,000 ppm CO$_2$. When multicellular life became common, CO$_2$ levels had come down under 20,000 ppm. When shellfish and corals first evolved, CO$_2$ levels were between 4,500 and 7,000 ppm.

There were many periods in the past when temperatures went up and CO$_2$ levels went down, and vice versa. The average CO$_2$ level for the last 600 million years was estimated at 2,300 ppm—nearly six times the current level.

Key: Average CO_2 has been approximately 2,300 ppm

Life sequesters CO_2. The levels of CO_2 in the atmosphere have been going down for millions of years. Some have estimated that CO_2 would have permanently dipped below the Lovelock Mass Extinction threshold of 150 ppm in about a million years. Humans have ruined that death and destruction scenario by returning CO_2 back into the atmosphere from whence it had come in the first place.

Today, crops benefit greatly from the increased CO_2. Yields have increased in part because of the CO_2 fertilization available to everyone on Earth. Scientists have found that crops in greenhouses with added CO_2, benefit with yield increases of 30–40% from 1,200 ppm CO_2. Some scientists have found that certain nutrients decrease in some crops, but the amount of decrease is minimal, while other nutrients increase. The net, overall effect is one of greater abundance in food and nutrition with more CO_2.

Investigators attempted to replicate the science which has raised concerns about more CO_2 in the oceans. What they found is that the alarm was based on bad science. Because CO_2 decreases the pH of water, unethical scientists had short-circuited their experiment by adding a different acid to achieve the desired pH. Ironically, when the pH change was achieved with CO_2, as it should have been, life thrived, perhaps because life is based on carbon.

Conclusion: We need more CO_2

Chapter 7: Climate Forecast

Warming alarmists are fond of telling us that we are doomed unless we take action now to reduce our "carbon" footprints. Ironically, carbon is the basis of all life on this planet. Eliminating carbon is equivalent to mass genocide.

But doomed? Hardly! We currently live in an Ice Age, near the bottom of Earth's livable temperature range. Warming is a good thing. If warming were to continue, we might end the Ice Age, people would have to move from the coasts, but life would thrive because of the calmer weather, greater rain and more abundant gas of life—carbon dioxide.

But climate changes in cycles.

Anyone who ever saw Al Gore's *An Inconvenient Truth* would remember the huge graph he showed where CO_2 and temperature rose and fell in unison. The warming peaks over the span of his graph averaged about 101,000 years—90,000 years for each glacial period and 11,000 years for each interglacial.

What Gore didn't tell his audience is that the temperature changes came first, driving the changes in CO_2.

Warmer oceans outgas dissolved carbon dioxide, while cold oceans soak up the gas.

But the real clincher on Gore's graph was the sharp spike in CO_2 to our then current level, but without a commensurate spike in temperature.

Remember, the greenhouse effect runs on the notion of CO_2 trapping infrared radiation. And that radiation travels at the speed of light, so it should not take very long to build up temperatures from increased carbon dioxide. The fact that we don't have the appropriate spike in temperature means that *CO$_2$ does not drive temperature!*

Key: Climate changes in cycles

Something else is driving global temperatures—not CO_2. Because of this, industry is off the hook, and carbon dioxide is off the hook. Heck, even global warming is off the hook. Not guilty of any climate crimes! All of them!

There are many cycles in climate. Overlapping cycles make the climate more complex and the resulting conditions harder to predict. But because CO_2 does *not* drive global temperature, we know that the current cycles are leading toward greater global cooling in the near future.

On the 1,000-year cycle of Holocene warm periods, we are likely at the peak of the Modern Warm Period and on our way into the next Little Ice Age, which may be slightly cooler than the last. In another thousand years, we will likely have another warm period, but its peak may be slightly cooler than that of the Modern.

After the far warmer Holocene Optimum, our world experienced a similar pattern of cooling, interrupted by the Minoan Warm Period. Could another spike of warmth interrupt the current, 3,000-year cooling trend? We don't know. We still don't know enough about all the factors which control our climate.

The factors which gave us the 100,000-year cycle of glacials and interglacials for the last million or so years may continue and give us an end to the Holocene in the near future (a few thousand years), or some other, unknown factor could trigger a different cycle, like the switch from 41,000 years to 100,000 years which occurred about 1.1 million years ago.

To cope with the coming cold, humanity will eventually need to come up with better methods of desalination, because there will be far less rain during a glacial period. But we still have plenty of time. Also, we will likely need to move our ports as sea level starts to fall again.

On the other hand, if we get lucky and figure out how to end the current Ice Age, then we will need to move cities to higher ground. But, again, we have plenty of time to take care of such gargantuan tasks.

The sub-title of this book states: "Nothing to fear." Worry is always a waste of energy. So is dwelling on fear. Human ingenuity has continued to solve problems, feeding an ever-increasing population that would have shocked Malthus —the original population alarmist. The fear mongers want us to do something in a panic. And that rarely, if ever, results in sound decisions for our future.

Global warming in an Ice Age is a good thing. But the fact that we've had a strong cooling trend for 3,000 years should not cause us any fear. Climate changes slowly. Despite the haranguing of the corrupt corporate media, our modern changes have been far slower and far more shallow than climate changes of the past. Even the insanity of governmental organizations should not cause us fear. If we take care of our own emotional health, give up our self-concern and rise to the level of responsibility to take care of our fellow Man, then we will have done all that we can do. We should rejoice at that and be humble to the truths as we discover them.

If we remain strong in the face of adversity, then we will have a chance to keep civilization thriving despite the mismanagement of some. We will either prepare civilization to weather the coming 90,000-year glacial period, or we will find techniques to end the current Ice Age and usher in a new epoch of prosperity.

The most important thing is how we treat each other, no matter what happens in our environment.

Conclusion: We need to end the current Ice Age or prepare for the coming cold

Appendix

- References
- Glossary
- Videography
- About Rod Martin, Jr.
- Other Books
- Connect

Illustrations

To view the illustrations in color use the following internet address:

https://tharsishighlands.wordpress.com/2018/11/18/illustrations-used-in-climate-basics/

References

Adam, David. (2007:1011). "Gore's climate film has scientific errors - judge." Retrieved on 2016:0116 from http://theguardian.com/environment/2007/oct/11/climat echange

Alley, Richard. (2000). "Holocene Epoch: Subatlantic Chronozone." Retrieved on 2016:0613 from http://s90.photobucket.com/user/dhm1353/media/Clim ate%20Change/Subatlantic_Had.png.html

Archive.org. (2012:0127). "Temperature -vs- CO2: Last 800,000 Years." Retrieved on 2016:0618 from http://web.archive.org/web/20120127155937/http://robe rtb.darkhorizons.org/TempGr/Vostok.JPG

Ball, Dr. Tim. (2015:1127). "An Important Lesson On The Anniversary of Climategate." Retrieved on 2016:0712 from https://wattsupwiththat.com/2015/11/27/an-important-lesson-on-the-anniversary-of-climategate/

Bastasch, Michael. (2015:0507). "NASA Warns About High CO2 Levels That Are Greening The Planet." Retrieved on 2016:0124 from http://dailycaller.com/2015/05/07/nasa-warns-about-high-co2-levels-that-are-greening-the-planet/

Bastasch, Michael. (2015:1217). "EXCLUSIVE: NOAA Relies On 'Compromised' Thermometers That Inflate US Warming Trend." Retrieved on 2016:1104 from http://dailycaller.com/2015/12/17/exclusive-noaa-relies-on-compromised-thermometers-that-inflate-u-s-warming-trend/

Bazell, Robert. (2010:1001). "U.S. apologizes for Guatemala STD experiments." Retrieved on August 8, 2016 from http://nbcnews.com/id/39456324/ns/health-sexual_health/t/us-apologizes-guatemala-std-experiments/

BBC. (2009:0210). "HD: Polar Bear on Thin Ice - Nature's Great Events: The Great Melt - BBC One." Retrieved on 2016:1021 from https://youtube.com/watch?v=Kv9v9ALV3yk

Becker, Jo, and Shane, Scott. (2012:0529). "Secret 'Kill List' Proves a Test of Obama's Principles and Will." Retrieved on 2016:1026 from http://nytimes.com/2012/05/29/world/obamas-leadership-in-war-on-al-qaeda.html

Berko, J.; Ingram, D.; *et al.* (2014:0730). "Deaths Attributed to Heat, Cold, and Other Weather Events in the United States, 2006–2010." Retrieved on 2015:1112 from http://cdc.gov/nchs/data/nhsr/nhsr076.pdf

Booker, Christopher. (2015:0301). "Dr Rajendra Pachauri: the clown of climate change has gone." Retrieved on 2016:0712 from http://telegraph.co.uk/comment/11441697/Dr-Rajendra-Pachauri-the-clown-of-climate-change-has-gone.html

Booker, Christopher. (2015:0207). "The fiddling with temperature data is the biggest science scandal ever."

Retrieved on 2016:1104 from
http://telegraph.co.uk/news/earth/environment/globalw
arming/11395516/The-fiddling-with-temperature-data-
is-the-biggest-science-scandal-ever.html

Brandt, Allan M. (1978:12). "Racism and Research: The Case of
the Tuskegee Syphilis Study." Retrieved on August 8,
2016 from
http://med.navy.mil/bumed/Documents/Healthcare%20
Ethics/Racism-And-Research.pdf

British Channel 4. (2007:0308). "The Great Global Warming
Swindle." Retrieved 2014:0830 from
https://youtube.com/watch?v=52Mx0_8YEtg

Broecker, W.S. (1998). "The End of the Present Interglacial:
How and When?" Quaternary Science Reviews, Vol. 17,
pp. 689-694, 1998.

Bunch, Will. (2005:0831). "Why the Levee Broke." Retrieved
on 2015:1113 from
http://alternet.org/story/24871/why_the_levee_broke

CBC. (ND). "The Ice Storm of 1998." Retrieved on 2016:1026
from http://cbc.ca/archives/topic/the-ice-storm-of-1998

Cohen, Tamara. (2013:0919). "World's top climate scientists
told to 'cover up' the fact that the Earth's temperature
hasn't risen for the last 15 years." Retrieved on
2016:1104 from http://dailymail.co.uk/news/article-
2425775/Climate-scientists-told-cover-fact-Earths-
temperature-risen-15-years.html

Connolly, Ronan, and Connolly, Michael. (2013:1205).
"Summary: 'Urbanization bias' — Papers 1–3."
Retrieved on 2016:1104 from
http://globalwarmingsolved.com/2013/12/summary-
urbanization-bias-papers-1-3/

Corbett Report. (2009:1121). "Climategate: Dr. Tim Ball on the hacked CRU emails." Retrieved on 2016:0708 from https://youtube.com/watch?v=Ydo2Mwnwpac

Corbett Report. (2013:0927). "The IPCC Exposed." Retrieved on 2016:0709 from https://youtube.com/watch?v=LOyBfihjQvI

Corbett, James. (2013:1203). "Genetic Fallacy: How Monsanto Silences Scientific Dissent." Retrieved on 2014:0108 from https://youtube.com/watch?v=ShJTcIlTna0

Corbett, James. (2016:0131). "Meet Maurice Strong: Globalist, Oiligarch, "Environmentalist." Retrieved on 2016:0709 from https://corbettreport.com/meet-maurice-strong-globalist-oiligarch-environmentalist/

Council on Foreign Relations. (2016:0629). "A Conversation With John O. Brennan." Retrieved on 2016:0709 from https://youtube.com/watch?v=uIQDqxl9FtM

CropsReview.com. (ND). "Plant Types: I. C3 Plants, Comparison with C4 and CAM Plants." Retrieved on 2016:1028 from http://cropsreview.com/c3-plants.html

CropsReview.com. (ND). "Plant Types: II. C4 Plants, Examples, and C4 Families." Retrieved on 2016:1028 from http://cropsreview.com/c4-plants.html

CropsReview.com. (ND). "Plant Types: III. CAM Plants, Examples and Plant Families." Retrieved on 2016:1028 from http://cropsreview.com/cam-plants.html

D'Aleo, Joseph. (2009:1013). "How bad is the global temperature data?" Retrieved on 2016:1104 from https://wattsupwiththat.com/2009/10/13/how-bad-is-the-global-temperature-data/

Deming, Dr. David. (2006:1206). "Statement of Dr. David Deming." Retrieved on 2016:0708 from http://epw.senate.gov/hearing_statements.cfm?id=2665

EngineeringToolbox.com. (ND). "Carbon Dioxide
Concentration - Comfort Levels." Retrieved on
2016:1026 from http://engineeringtoolbox.com/co2-
comfort-level-d_1024.html

Fairbanks, R. (1989:1207). "A 17,000-year glacio-eustatic sea
level record: influence of glacial melting rates on the
Younger Dryas event and deep-ocean circulation,"
Nature, Vol. 342.

Georgia State University. (ND). "Systems of Photosynthesis."
Retrieved on 2016:1026 from http://hyperphysics.phy-
astr.gsu.edu/hbase/biology/phoc.html

Gerhart, Laci M., and Ward, Joy K. (2010:0705). "Plant
responses to low [CO2] of the past." Retrieved on
2016:1027 from
http://onlinelibrary.wiley.com/doi/10.1111/j.1469-
8137.2010.03441.x/pdf

Goddard, Steven. (2012:0712). "1988 : James Hansen And Tim
Wirth Sabotaged The Air Conditioning In Congress."
Retrieved on 2016:0618 from
https://stevengoddard.wordpress.com/2012/07/12/1988-
james-hansen-and-tim-wirth-sabotaged-the-air-
conditioning-in-congress/

Goldenberg, Suzanne. (2014:0922). "Heirs to Rockefeller oil
fortune divest from fossil fuels over climate change."
Retrieved on 2016:0127 from
http://theguardian.com/environment/2014/sep/22/rocke
feller-heirs-divest-fossil-fuels-climate-change

GreenLearning. (2009:0625). "1 MILLION pounds of Food on 3
acres. 10,000 fish 500 yards compost." Retrieved on
2015:1113 from
https://youtube.com/watch?v=jV9CCxdkOng

Griffin, G. Edward. (1994). *The Creature from Jekyll Island: A Second Look at the Federal Reserve.* American Media: Westlake Village, California

Grunwald, Michael. (2005:0908). "Money Flowed to Questionable Projects." Retrieved on 2015:1113 from http://washingtonpost.com/wp-dyn/content/article/2005/09/07/AR2005090702462.html

Haby, Jeff. (ND). "What Causes the Wind to Blow?" Retrieved on 2016:1103 from http://theweatherprediction.com/basic_weather_questions/wind.html

Hasegawa, H., Tada, R., *et al.* (2012:0823). "Drastic shrinking of the Hadley circulation during the mid-Cretaceous Supergreenhouse." Retrieved on 2015:1119 from http://clim-past.net/8/1323/2012/cp-8-1323-2012.pdf

Heller, Tony. (2014:0623). "NOAA/NASA Dramatically Altered US Temperatures After The Year 2000." Retrieved on 2016:1104 from https://stevengoddard.wordpress.com/2014/06/23/noaa-nasa-dramatically-altered-us-temperatures-after-the-year-2000/

Humlum, Ole. (ND). "10,700 years — GISP2 — with CO2 from EPICA DOME C." Retrieved on 2016:0613 from http://climate4you.com/images/GISP2%20Temperature Since10700%20BP%20with%20CO2%20from%20EPICA%20DomeC.gif

James, John T., Macatangay, Ariel. (ND). "Carbon Dioxide - Our Common 'Enemy'." Retrieved on 2016:1029 from https://ntrs.nasa.gov/archive/nasa/casi.ntrs.nasa.gov/20090029352.pdf

Kollipara, Puneet. (2014:0122). "Earth Won't Die as Soon as Thought." Retrieved on 2016:1029 from http://sciencemag.org/news/2014/01/earth-wont-die-soon-thought

Lamb, Hubert H. (2012:1124). *Climate: Present, Past and Future: Volume 2: Climatic History and the Future.* Routledge Revivals, Abingdon-on-Thames, UK.

Law, Jennifer, Watkins, Sharmi, *et al.* (2010:06). "In-Flight Carbon Dioxide Exposures and Related Symptoms: Association, Susceptibility, and Operational Implications." Retrieved on 2016:1029 from https://ston.jsc.nasa.gov/collections/trs/_techrep/TP-2010-216126.pdf

LoGiurato, Brett. (2013:0625). "Obama's Climate Joke About A 'Flat Earth Society' Actually Referenced A Real Group." Retrieved on 2016:1025 from http://businessinsider.com/flat-earth-society-to-obama-climate-change-speech-georgetown-2013-6

Lovelock, J.E. & Whitfield, M. (1982:0408) "Lifespan of the Biosphere." *Nature* 296, 561–563.

Lüning, Sebastian. (2016:0109). "Evidence of the Medieval Warm Period in Australia, New Zealand and Oceania." Retrieved on 2016:0713 from https://wattsupwiththat.com/2016/01/09/evidence-of-the-medieval-warm-period-in-australia-new-zealand-and-oceania/

Manning, Katie, Timpson, Adrian. (2014:0701). "The demographic response to Holocene climate change in the Sahara." Retrieved on 2016:1103 from http://sciencedirect.com/science/article/pii/S0277379114002728

Martin, Jr., Rod. (2010:0509). "Atlantis Quest—Uncovering the Secrets that Prove Plato Right." Retrieved on 2010:0509 from http://hubpages.com/education/Atlantis-Quest-Uncovering-the-Secrets-that-Prove-Plato-Right

Martin, Jr., Rod. (2015:09). *Dirt Ordinary: Shining a Light on Conspiracies*. Tharsis Highlands: Cebu, Philippines

Martin, Jr., Rod. (2016:0719). "Top 10 Climate Change Lies Exposed." Retrieved on 2016:0719 from https://youtube.com/watch?v=ICGaI_8qI8c

Martin, Jr., Rod. (2016:0826). *Thermophobia: Shining a Light on Global Warming*. Tharsis Highlands, Cebu, Philippines.

Martin, Jr., Rod. (2016:11). *Red Line — Carbon Dioxide: How humans saved all life on Earth by burning fossil fuels.* Tharsis Highlands, Cebu, Philippines.

Maue, Ryan. (2014:0930). "Global Hurricane Frequency." Retrieved on 2015:1114 from http://policlimate.com/tropical/global_major_freq.png

Maue, Ryan. (2014:0930). "Global Tropical Cyclone Accumulated Cyclone Energy (ACE). Retrieved on 2015:1114 from http://policlimate.com/tropical/global_running_ace.png

McKitrick, Ross. (ND). "The Graph of Temperature vs. Number of Stations." Retrieved on 2016:1104 from http://uoguelph.ca/~rmckitri/research/nvst.html

Mortensen, Lars. (2004). "Doomsday Called Off." Retrieved on 2016:1107 from https://youtube.com/watch?v=Pg_2fJImqac [http://www.imdb.com/title/tt0493121/]

Mortensen, Lars. (2008). "The Cloud Mystery." Retrieved on 2015:1015 from https://youtube.com/watch?v=ANMTPF1blpQ [http://imdb.com/title/tt2005265/]

Mudelsee, M., et al, (2003:0911). "No upward trends in the occurrence of extreme floods in central Europe." Retrieved on 2016:0618 from http://nature.com/nature/journal/v425/n6954/full/nature 01928.html

Mudelsee, M., et al. (2004:1202). "Extreme floods in central Europe over the past 500 years: Role of cyclone pathway 'Zugstrasse Vb'." Retrieved on 2016:0618 from http://onlinelibrary.wiley.com/doi/10.1029/2004JD00503 4/full

NASA. (2013:0203). "Consensus: 97% of climate scientists agree." Retrieved on 2015:0728 from http://climate.nasa.gov/scientific-consensus/

NCDC. (ND). Graph: "Jan-Dec Global Mean Temperature over Land & Ocean." Retrieved 2015:1114 from http://ncdc.noaa.gov/sotc/service/global/global-land-ocean-mntp-anom/201001-201012.gif

NOAA. (ND). "U.S. Annual Count of EF-1+ Tornadoes, 1954 through 2014." Retrieved 2015:1029 from http://www1.ncdc.noaa.gov/pub/data/cmb/images/torn ado/clim/EF1-EF5.png

NOAA. (ND). "U.S. Annual Count of Strong to Violent Tornadoes (F3+), 1954 through 2014." Retrieved 2015:1029 from http://www1.ncdc.noaa.gov/pub/data/cmb/images/torn ado/clim/EF3-EF5.png

NOAA. (ND). "GISP2 Volcanic Markers." Retrieved on 2008:0801 from ftp://ftp.ncdc.noaa.gov/pub/data/paleo/icecore/greenlan d/summit/gisp2/chem/volcano.txt

Norris, Richard. (ND). "Cretaceous Thermal Maximum ~85-90 Ma." Retrieved on 2016:1103 from http://scrippsscholars.ucsd.edu/rnorris/book/cretaceous -thermal-maximum-85-90-ma

NSIDC. (ND). "Quick Facts on Ice Sheets." Retrieved on 2015:1113 from https://nsidc.org/cryosphere/quickfacts/icesheets.html

O'Connor, J.E., and Costa, J.E. (2004). "The World's Largest Floods, Past and Present: Their Causes and Magnitudes." Retrieved on 2016:0618 from http://pubs.usgs.gov/circ/2004/circ1254/pdf/circ1254.pdf

Painting, Rob. (ND). "Positives and negatives of global warming: Intermediate" Retrieved 2016:0611 from http://skepticalscience.com/argument.php?f=global-warming-positives-negatives&l=2

Paterson, W.S.B. (1972:11). "Laurentide Ice Sheet: Estimated Volumes during Late Wisconsin." Retrieved on 2015:1113 from http://onlinelibrary.wiley.com/doi/10.1029/RG010i004p00885/pdf

Phys.org. (2012:0614). "Warm climate -- cold Arctic? The Eemian is a poor analogue for current climate change." Retrieved on 2016:1026 from http://phys.org/news/2012-06-climate-cold-arctic-eemian.html

Pielke, Jr., Roger. (2014:0908). Tweet: "Phoenix floods, climate change!" Retrieved on 2016:0618 from https://twitter.com/RogerPielkeJr/status/509021248039313408/photo/1

Randerson, James. (2010:0127). "University in hacked climate change emails row broke FOI rules." Retrieved on 2016:0713 from https://theguardian.com/environment/2010/jan/27/uea-hacked-climate-emails-foi

Rose, David. (2015:0118). "Nasa climate scientists: We said 2014 was the warmest year on record... but we're only 38% sure we were right." Retrieved on 2016:0118 from http://dailymail.co.uk/news/article-2915061/Nasa-climate-scientists-said-2014-warmest-year-record-38-sure-right.html

Sage, Rowan F., and Zhu, Xin-Guang. (2011). "Exploiting the engine of C4 photosynthesis." Retrieved on 2016:1029 from http://jxb.oxfordjournals.org/content/62/9/2989.full.pdf

SciJinks. (2016:1018). "Why does wind blow?" Retrieved on 2016:1103 from http://scijinks.jpl.nasa.gov/wind/

Schneider, Mike. (2014:0103). "Florida Will Soon Have More People Than New York." Retrieved on 2015:1112 from http://businessinsider.com/florida-will-soon-have-more-people-than-new-york-2014-1

ScienceDaily.com. (2013:0708). "Deserts 'greening' from rising carbon dioxide: Green foliage boosted across the world's arid regions." Retrieved on 2016:0124 from http://sciencedaily.com/releases/2013/07/130708103521.htm

Solomon, Lawrence. (2011:0103). "Lawrence Solomon: 97% cooked stats." Retrieved on 2016:0127 from http://business.financialpost.com/fp-comment/lawrence-solomon-97-cooked-stats

Spencer, Dr. Roy. (ND). "Latest Global Temps." Retrieved on 2016:1104 from http://drroyspencer.com/latest-global-temperatures/

Swann, Ben. (2012:0904). "Reality Check: 1 on 1 With President Obama, How Does He Justify A Kill List?" Retrieved on 2012:0905 from https://youtube.com/watch?v=WrRuNOaNYME

Taylor, James. (2013:0710). "Global Warming? No, Satellites Show Carbon Dioxide Is Causing 'Global Greening'." Retrieved on 2016:0124 from http://forbes.com/sites/jamestaylor/2013/07/10/global-warming-no-satellites-show-carbon-dioxide-is-causing-global-greening/

University of Copenhagen. (ND). "A glimpse into the
      Eemian." Retrieved on 2015:0807 from
      http://iceandclimate.nbi.ku.dk/research/climatechange/
      glacial_interglacial/eemian/

Wakefield, Dr. Andrew. (2016:0401). *Vaxxed: From Cover-Up to
      Catastrophe*. Cinema Libre Studio: Burbank, California

Wall Street Journal. (2009:1124). "Global Warming With the
      Lid Off. The emails that reveal an effort to hide the
      truth about climate science." Retrieved on 2016:0713
      from
      http://wsj.com/articles/SB10001424052748704888404574
      547730924988354

Watts, Anthony. (ND). "Weather Stations." Retrieved on
      2016:1104 from
      https://wattsupwiththat.com/category/weather_stations
      /

Watts, Anthony. (2009:1122). "CRU Emails 'may' be open to
      interpretation, but commented code by the
      programmer tells the real story." Retrieved on
      2016:0708 from
      https://wattsupwiththat.com/2009/11/22/cru-emails-
      may-be-open-to-interpretation-but-commented-code-
      by-the-programmer-tells-the-real-story/

Watts, Anthony. (2009:1122). "Bishop Hill's compendium of
      CRU email issues." Retrieved on 2016:0708 from
      https://wattsupwiththat.com/2009/11/22/bishop-hills-
      compendium-of-cru-email-issues/

Watts, Anthony. (2012:0730). "Surface Stations, A resource for
      climate station records and surveys." Retrieved on
      2016:1104 from http://surfacestations.org/

Watts, Anthony. (2013:0411). "Evidence for a Global Medieval
      Warm Period." Retrieved on 2016:0713 from

https://wattsupwiththat.com/2013/04/11/evidence-for-a-global-medieval-warm-period/

Watts, Anthony. (2013:0827). "Cook's 97% climate consensus paper crumbles upon examination." Retrieved on 2016:0127 from http://wattsupwiththat.com/2013/08/28/cooks-97-climate-consensus-paper-crumbles-upon-examination/

Watts, Anthony. (2013:1014). "90 climate model projectons versus reality." Retrieved on 2016:1102 from https://wattsupwiththat.com/2013/10/14/90-climate-model-projectons-versus-reality/

Watts, Anthony. (2013:1208). "The truth about 'We have to get rid of the medieval warm period'." Retrieved on 2016:0708 from https://wattsupwiththat.com/2013/12/08/the-truth-about-we-have-to-get-rid-of-the-medieval-warm-period/

Watts, Anthony. (2014:0328). "IPCC admission from new report: 'no evidence climate change has led to even a single species becoming extinct'." Retrieved on 2016:1102 from https://wattsupwiththat.com/2014/03/28/ipcc-admission-from-new-report-no-evidence-climate-change-has-led-to-even-a-single-species-becoming-extinct/

Watts, Anthony. (2014:0908). "Phoenix flooding – not due to 'climate change', extreme rainfall events are not on the increase." Retrieved on 2016:0618 from https://wattsupwiththat.com/2014/09/08/phoenix-flooding-not-due-to-climate-change-extreme-rainfall-events-are-not-on-the-increase/

Watts, Anthony. (2014:1112). "Kashmir Floods Nothing New, Not Due To Climate Change." Retrieved on 2016:0618 from https://wattsupwiththat.com/2014/11/12/kashmir-floods-nothing-new-not-due-to-climate-change/

Watts, Anthony. (2015:0403). "Claim: polar bears can't subsist on anything but seals." Retrieved on 2016:1028 from https://wattsupwiththat.com/2015/04/03/claim-polar-bears-cant-subsist-on-anything-but-seals/

Watts, Anthony. (2015:1010). "USGS puts the kibosh on '1000 year flood' and 'caused by climate change' claims over South Carolina flooding." Retrieved on 2016:0618 from https://wattsupwiththat.com/2015/10/10/usgs-puts-the-kibosh-on-1000-year-flood-and-caused-by-climate-change-claims-over-south-carolina-flooding/

Watts, Anthony. (2016:0206). "Record Missouri flooding was manmade calamity, not climate change, scientist says." Retrieved on 2016:0618 from https://wattsupwiththat.com/2016/02/06/record-missouri-flooding-was-manmade-calamity-not-climate-change-scientist-says/

WattsUpWithThat.com. (2015:0120). "2014 The Most Dishonest Year On Record." Retrieved on 2016:0119 from http://wattsupwiththat.com/2015/01/20/2014-the-most-dishonest-year-on-record/

Waugh, Rob. (2011:1128). "Climategate scientists DID collude with government officials to hide research that didn't fit their apocalyptic global warming." Retrieved on 2016:0713 from http://dailymail.co.uk/sciencetech/article-2066240/Second-leak-climate-emails-Political-giants-weigh-bias-scientists-bowing-financial-pressure-sponsors.html

WeatherWizKids.com. (ND). "Wind." Retrieved on 2016:1103 from http://weatherwizkids.com/weather-wind.htm

Wikipedia.org. (ND). "January 1998 North American ice
        storm." Retrieved on 2016:1026 from
        https://en.wikipedia.org/wiki/January_1998_North_Am
        erican_ice_storm

Glossary

<u>Note:</u> Not every term or concept has been included in this glossary. I encourage you to explore the subject online or in books on those terms for which you would like more information. Make learning a lifetime occupation.

carbon dioxide *n.*—an odorless, colorless gas and a minor constituent of the Earth's atmosphere. Without this trace gas, all life on Earth would die. Frequently abbreviated CO_2. This is what plants breathe. And plants "exhale" oxygen (which see). Not to be confused with poisonous carbon monoxide (CO).

climate *n.*—a persistent average state of a region's weather, typically taken over a period of several decades—usually thirty years.

climate change *n.*—modification of a region's persisting average weather. This can include warming or cooling, alterations in turbulence, patterns of flow, timing of events, atmospheric chemistry and more. Such modifications have occurred throughout the existence of our planet's atmosphere—more than 4 billion years. This term has been kidnapped by a modern movement to mean only "recent, manmade, warming and cata-strophic" modifications to the atmosphere. This is a distortion of the original definition.

CO_2 *n.*—carbon dioxide (which see).

drought *n.*—a period of decreased rainfall that is insufficient for the life forms within a region. Drought typically occurs from changes in weather patterns, but more importantly from regional or global cooling. In fact, the global cooling of the last 50 million years or so has significantly desiccated the planet, increasing the extent of subtropical deserts and creating polar deserts.

Earth *n.*—our home planet. It possesses a breathable atmosphere, water in three key phases (solid, liquid, gas), dry land and a surface teeming with life. It also has one natural satellite typically called the Moon.

glacial *n.*—a cooler period of increased glaciation during an Ice Age in which polar glaciers expand to cover large portions of adjacent continents. During the last 1.1 million years of the current Ice Age, glacial periods have averaged 90,000 years in length (ref: W.S. Broecker, 1998). The duration of glacial periods for the last 800,000 years has varied between 24,000 and 143,000 years. Compare *interglacial.*

global cooling *n.*—a decrease in the average temperature of the planet. This can be a bad thing during our current Ice Age. Cooling tends to produce less evaporation, and thus drier climate.

global warming *n.*—an increase in the average temperature of the planet. This can be a good thing during our current Ice Age. Warming tends to produce more evaporation, and thus moister climate.

Holocene *n.*—an interglacial of the current Ice Age; the current interglacial (which see).

Holocene Optimum *n.*—a period of about 3,000 years wherein the northern hemisphere was as much as 1.1°C warmer than today. This warmth, compared with the shallow cool periods (roughly as warm as today) resulted in a green Sahara for about 3,000 years.

hurricane *n.*—A dangerous tropical cyclone of the Atlantic Ocean region. Compare *typhoon*.

Ice Age *n.*—a period of increased cooling where both polar regions experience permanent glaciation throughout the year. The current such period has had glaciation in Greenland and Antarctica for roughly 2.6 million years. Such periods include several glacial and interglacial periods, alternating between warmer and cooler phases.

interglacial *n.*—a warmer period of relaxed glaciation during an Ice Age in which polar glaciation recedes and global climate warms noticeably. The amount of warming and glacial receding can vary a great deal with some such periods being as much as 5°C warmer than our current Modern Warm Period, or 2°C cooler. There have been several dozen interglacials in the current Ice Age. For the last 1.1 million years, interglacials have averaged about 11,000 years in length (ref: W.S. Broecker, 1998). The duration of interglacial periods for the last 800,000 years has varied between 4,000 and 24,000 years. Compare *glacial*.

Intergovernmental Panel on Climate Change *n.*—a political organization associated with the United Nations tasked with determining the nature and extent of man's impact on the planetary climate as a result of burning fossil fuels.

IPCC *n.*—Intergovernmental Panel on Climate Change (which see).

Jupiter *n.*—the largest planet in our star system, roughly ten times the diameter of Earth, with a thick atmosphere many thousands of kilometers deep and no solid surface. Because of its great distance from the sun, it is extremely cold at the tops of the clouds, near the air pressure equivalent to that on Earth's surface. Despite the extreme cold, the planet hosts some of the largest storms in the solar system.

oxygen *n.*—a key constituent of Earth's atmosphere and the most vital gas for animal life. Animals exhale carbon dioxide (which see).

parts per million *n.*—the concentration of something as a fractional measure compared to a whole. If you take a million of something, the count given will be the number of pieces or portions out of that whole million that apply to a specific substance. This is similar to the term percent. Example: The atmosphere consists of 21 percent oxygen, or 210,000 parts per million oxygen.

Pleistocene *n.*—the current Ice Age (which see). This period of permanent polar glaciation has lasted for 2.6 million years. Before scientists knew very much about Earth's history, they thought the Pleistocene ended 11,600 years ago. Today, we know that the current epoch—the Holocene—is merely one in a series of dozens of interglacial periods that are part of this Ice Age.

ppm *abbr.*—parts per million (which see).

tornado *n.*—a small, cyclonic storm, typically less than several hundred meters across, with extremely fast winds and an ability to create tremendous damage to buildings and to anything else above ground.

typhoon *n.*—a dangerous tropical cyclone of the Pacific Ocean region. Compare *hurricane*.

Venus *n.*—our sister planet, closer to the sun. The planet is slightly smaller than Earth (which see), has a crushing atmosphere of mostly CO_2, a heavily reflective cloud cover, and a surface with virtually no wind and temperatures hot enough to melt lead (462°C). The planet spins very, very slowly and has no natural satellite.

warm period *n.*—a span of time which has a higher temperature than the preceding and succeeding spans of time. Climate always changes and most frequently in repeating cycles. The Holocene has contained 10 clearly-defined major warm periods on a roughly 1,000-year cycle. Cycles of other periods make the pattern of warming and cooling more complex than they would be if there were only one cycle involved. The most recent four major warm periods of the Holocene have been, the Modern (1850 to today), the Medieval (850–1350), the Roman (200 BC–AD 100) and the Minoan (1400–1100 BC).

Videography

Be sure to Like, Comment and Subscribe to the channel:
https://youtube.com/c/RodMartinJr/

Climate Change Lies Exposed series

Top 10 Climate Change Lies Exposed
https://youtube.com/watch?v=ICGal_8qI8c
Climate Change Lie #1 Exposed: Global Warming is Bad
https://youtube.com/watch?v=KbfjEPo083U
Climate Change Lie #2 Exposed: CO2 Causes Dangerous
Global Warming
https://youtube.com/watch?v=ZH5ATcpMJQo
Climate Change Lie #3 Exposed: Global Warming Causes
Extreme Weather
https://youtube.com/watch?v=aTiBbAGl0qI
Climate Change Lie #4 Exposed: Global Warming causes
droughts
https://youtube.com/watch?v=DusZ5dP4hDw
Climate Change Lie #5 Exposed: Our current warmth is
unusual
https://youtube.com/watch?v=FR2aZc5bjUU

Climate Change Lie #6 Exposed: Our current level of CO2 is unusually high
https://youtube.com/watch?v=ASV3UUwYZg0
Climate Change Lie #7 Exposed: The rate of warming is dangerous
https://youtube.com/watch?v=OsJ67Hp4l-g
Climate Change Lie #8 Exposed: The Science is Settled
https://youtube.com/watch?v=6yzkAjWY8rM
Climate Change Lie #9 Exposed — There is a consensus on dangerous, man made, Global Warming
https://youtube.com/watch?v=URE4NMk1DbA

Carbon Dioxide Fan Club

Earth vs. Venus: Will our world ever suffer runaway greenhouse warming?
https://youtube.com/watch?v=SO1M8GEDyYk
Top 10 Facts that Prove CO2 Does NOT Drive Global Temperature
https://youtube.com/watch?v=CSQlJx76b64
Verdict: CO2 Not Guilty! Greenhouse DESTROYED! Must see!
https://youtube.com/watch?v=1f6zB320Hac

Global Warming Fan Club

How Global Warming Made Civilization Possible
https://youtube.com/watch?v=057GgxpZWRc
Top 10 Reasons Global Warming is Good
https://youtube.com/watch?v=dQc4iXgrrEo

Big Climate Quiz (BCQ)

BCQ #1: Why didn't civilization start during the last glacial period?
https://youtube.com/watch?v=Bf0gty2XAjw
BCQ #2: What Causes Wind to Blow?

https://youtube.com/watch?v=lhk7JIQ6e-U
BCQ #3: How does land ever get water?
https://youtube.com/watch?v=do0kb7Udq-g
BCQ #4: What is an Ice Age?
https://youtube.com/watch?v=RjMbE-G8JFo

Climate Music Video series

Thermophobia - Why Fear of Warming in the current Ice Age is all wrong
https://youtube.com/watch?v=Q68fIkdC9Rk
Extreme Weather - How the Climate Change Alarm is All Wrong
https://youtube.com/watch?v=x18gwLpLI2A
Thermophobia -- Debunking: "Global Warming causes more storms"
https://youtube.com/watch?v=d40_2yGuV_o

About Rod Martin, Jr.

Rod Martin, Jr. is a modern polymath (Renaissance man)—artist, scientist, mathematician, engineer and philosopher. He first became interested in climate science in the mid-70s. A forest ecology PhD friend of his was retiring and donated two climate texts to the cause. Initially, Martin's interest in the subject covered planetary atmospheres—weather systems, atmospheric retention rates, optical thickness (greenhouse effect), adiabatic lapse rates, climate chemistry and planetary habitability.

Like so many others, during the 70s, 80s and 90s, Martin's interest in ecology and the environment continued to grow. When Al Gore's film, An Inconvenient Truth, came out in 2006, Martin was an immediate fan. But as the controversy on the topic heated up, Martin suddenly realized that all of the things he had learned about climate over the years contradicted many of the so-called facts in Gore's award-winning film.

In college, from the mid-90s to the early 2000s, Martin studied computer science, earning a degree, summa cum laude.

With only a 139 IQ, Martin realized that he was not the sharpest implement in the tool shed. In fact, all of his younger brothers had far higher IQs. From this relative handicap, he learned the immense value of humility and the need to remain unattached to any ideas, lest they become dogma, and blind him from further discovery. Thus, he was able to learn the true value of skepticism, and was able to recognize the inevitable pitfalls of that scientific paradigm. He also made the distinction between confidence in knowledge (an enormous source of blindness) and confidence in one's ability to find new knowledge (a source of empowerment).

In 2016, Martin implemented a campaign to set the record straight on climate. He wasn't alone. Many climate scientists, astrophysicists, meteorologists and concerned citizens had already begun to speak out against the so-called "scientific consensus" (an oxymoron, because science is never done by consensus). Martin has created numerous educational videos on climate change and global warming, and created a website to discuss these topics in greater detail.
https://GlobalWarmthBlog.WordPress.com/

From a lasting love of stars and astronomy, he created 3D space software, "Stars in the NeighborHood," available online.
https://SpaceSoftware.WordPress.com/buy-now/

He currently resides in the Philippines with his wife, Juvy.

He has taught mathematics, information technology, critical thinking and professional ethics at Benedicto College, Mandaue City, Cebu. He continues to teach online and to write.

Other Books by Rod Martin, Jr.

Non-Fiction (as Rod Martin, Jr.)

The Art of Forgiveness, Tharsis Highlands (2012, 2015)

The Bible's Hidden Wisdom: God's Reason for Noah's Flood, Tharsis Highlands (2014)

The Spark of Creativity, Tharsis Highlands (2014)

Dirt Ordinary: Shining a Light on Conspiracies, Tharsis Highlands (2015)

Favorable Incompetence: Shining a Light on 9/11, Tharsis Highlands (2015)

Thermophobia: Shining a Light on Global Warming, Tharsis Highlands (2016)

Red Line — Carbon Dioxide: How humans saved all life on Earth by burning fossil fuels, Tharsis Highlands (2016)

The Science of Miracles: How Scientific Method Can Be Applied to Spiritual Phenomena, Tharsis Highlands (2018)

Proof of God, Tharsis Highlands (2018)

Deserts & Droughts: How does Land Ever Get Water?, Tharsis Highlands (2018)

Taking Charge: How to Assert Positive Control Over Your Own Emotions, Tharsis Highlands (2018)

Spirit is Digital—Science is Analog: Discovering where miracles and logic intersect, Tharsis Highlands (2019)

Proof of Atlantis? Evidence of Plato's Lost Island Empire, Tharsis Highlands (2019)

Enemies of Christ: The Need to Protect Our Own Salvation from Ravening Wolves, Tharsis Highlands (2019)

Science Fiction (as Carl Martin)

Touch the Stars: Emergence, with John Dalmas, Tor (1983), *expanded* Tharsis Highlands (2012)

Touch the Stars: Diaspora, Book 2 of Touch the Stars, Tharsis Highlands (2014)

Entropy's Children, anthology of short fiction, Tharsis Highlands (2014)

Gods and Dragons, Book 1 of *Edge of Remembrance*, Tharsis Highlands (2017)

Tales of Atlantis Lost, Book 2 of *Edge of Remembrance*, Tharsis Highlands (2017)

An Excerpt from *Thermophobia*

Chapter 1: Unreasonable Fear

"When the wise man points at the moon, the fool looks at the finger." —Attributed to be a Chinese proverb

What you don't know about global warming could kill you. The only problem is, the United Nations, NASA and so many other corrupt bodies have it upside-down and backwards. Listening to them will pollute your mind.

It may sound ludicrous to say that the world is wrong and little old me is right. But let us use the facts, rather than popularity or authority to rule our judgment. As you will learn, it's not the "world" in opposition to this work. No, it's a far, far smaller portion who lie about their numbers. To them, perception is everything, and if you perceive only the lie, then that lie becomes your reality. Four years ago, that was me.

First of all, an unreasonable fear, not based upon facts, is typically called *paranoia*. This is a byproduct of the current epidemic of thermophobia. Fear of warmth is a manufactured fear. As we all know, no one in their right mind fears cuddling with a healthy loved one. No one is terrified of a warm shower or a hot meal.

These days, when something bad happens in the weather, the mainstream news media points to global warming as the cause. Heat wave: global warming. Cold snap: global warming. Hurricanes and tornadoes: global warming. Drought: global warming. Flood: global warming. Economic upheaval: global warming. Election lost to the Democrats: global warming. Election lost to the Republicans: global warming. And since global warming took a holiday after 1998, simply replace the term with "climate change," but keep the Carbon Tax, because we need to reduce the global warming which no longer exists. Confused? Could this be by design?

The problem is, they've come up with short, but inaccurate terms that are misleading. By "global warming," they mean "man made catastrophic global warming." But the ironic thing is, there is very little truth in what they say. They confuse real damage from chemical pollution with fictional damage from warming and CO2. Nearly all of the disasters about which they wring their hands are false. By the time you're finished with this book, you will understand *why* they are false.

https://tharsishighlands.wordpress.com/books/thermophobi a-global-warming/

Connect with Rod Martin, Jr.

Rod Martin, Jr. is his pen name for non-fiction. Carl Martin is his pen name for fiction.
BitChute—https://bitchute.com/channel/M63WrjRpNSPT/
Minds—https://minds.com/RodMartinJr
Gab—https://gab.ai/RodMartinJr
Website and Blog—https://rodmartinjr.wordpress.com/
HubPages—https://hubpages.com/@lone77star
Smashwords author page—
        https://smashwords.com/profile/view/CarlMartin77
Smashwords author page—
        https://smashwords.com/profile/view/RodMartinJr
Udemy courses page—https://udemy.com/user/rodmartinjr/
Facebook—https://facebook.com/RodMartinJr/
Twitter—https://twitter.com/LoneStar77/
Google+—https://plus.google.com/+RodMartinJr/
YouTube—https://youtube.com/c/RodMartinJr/
Goodreads author page—https://goodreads.com/Carl_Martin
Goodreads author page—
        https://goodreads.com/Rod_Martin_Jr

Amazon author page—https://amazon.com/Carl-
    Martin/e/B008CX8KN6/
Amazon author page—https://amazon.com/Rod-Martin-
    Jr/e/B008CZ9JTS/

Made in the USA
Coppell, TX
05 May 2020